Butterfly Wings

Snow Canyon Publishing
All Rights Reserved

Bonnie the butterfly is feeling a little plump today, not because she has been feasting on the delicious, sweet nectar from the beautiful flowers. It's because she is ready to lay her spring eggs!

Unlike caterpillars who live on a diet of green leaves, adult butterflies feed only on liquid nectar found in flowers. The adult butterfly drinks the sweet nectar through a tube-like tongue called a proboscis. The tongue uncoils to sip the nectar and then coils up again when the butterfly has finished feeding.

Bonnie gently places her eggs on the perfect leaf that is protected from the rain and the sun.

A female butterfly will lay her eggs as soon as four days after emerging from her chrysalis as an adult. She is very selective about where to lay her eggs. She instinctively knows which plants will serve as suitable food for her hungry little caterpillars when they hatch. A sticky substance produced by the female enables the eggs to stick wherever she lays them on the underside of a leaf or on a stem.

After seven long days of waiting, Bonnie anxiously watches as her eggs hatch. Out of the eggs climb cute, little caterpillars. Her babies do not yet look like her with her colorful wings and a body that shimmers in the sunlight.

Some butterflies lay a single egg, while others may lay their eggs in clusters. If you look close enough you can actually see the tiny caterpillar growing inside of the egg. There are four stages in the life cycle of the butterfly, with **stage one** being the egg, or "ovum."

Bonnie smiles proudly at her little caterpillars. They look happily back at their mother. She is so beautiful!

Fortunately the butterfly will lay her eggs on the type of leaf or plant her little caterpillars can eat. When the eggs hatch, the young caterpillars immediately get to work eating the leaf or plant they hatched on. Each species of caterpillar prefers a different type of plant. Because they are so small and unable to cover large distances, caterpillars may spend their entire lives on the same plant or leaf!

As the little caterpillars play on the spring flowers and feast upon the fresh green leaves, some butterflies from a nearby tree fly past the little caterpillars. The other butterflies laugh and point at the little caterpillars. One butterfly shouts out, "You are not as beautiful as we are! You don't even have wings."
The other butterflies snicker at the little caterpillars.

The larva, or caterpillar, that hatches from the egg is **stage two** in the butterfly life cycle.

Sad and ashamed, the little caterpillars crawl home to their mother.
"Why the sad tears?" asks their mother.

The primary goal of the caterpillar is to eat as much as they can to grow large enough to begin the chrysalis (pupate) stage. Caterpillars have chewing mouth parts, called mandibles, which enable them to eat leaves and other plant parts. This is very different from the tube-like tongue that the newly formed butterfly will have when it emerges from its chrysalis. During the chrysalis stage, the caterpillar's mouth will change and the tube-like tongue will form. Nectar will then become its total source of food.

"We are not beautiful like you and the butterflies in the other trees," cried the little caterpillars. "We don't even have wings. And we don't glisten in the sunlight like they do."

Other than the plants and leaves that the caterpillars eat, they do not need to drink additional water. All of the water they need to sustain their developing body is found within the plants and leaves.

"Don't worry," Bonnie reassures her little caterpillars. "One day you too will grow your wings, and you will also glisten in the sunlight just like me and the other butterflies. The other butterflies do not see you for *who* you really are, and for *what* you will become. Pay no attention to them."

A caterpillar's primary activity is eating. They have a large appetite and eat constantly. As the caterpillar continues to eat, its body grows considerably. The tough outer skin, or exoskeleton, however, does not grow or stretch along with the enlarging caterpillar. Instead, the old exoskeleton is shed in a process called molting and it is replaced by a new, larger exoskeleton. A caterpillar may go through as many as four to five molts before it begins the chrysalis process.

The next day Bonnie tells to her little caterpillars that it is time for them to take a long nap. She explains that their skin will soon change into a soft, warm blanket that will keep them safe while they sleep.

Stage three in the butterfly life cycle is known as the chrysalis or pupa stage. Though often used interchangeably, a chrysalis and a cocoon are not the same. A cocoon is "spun" around itself by a caterpillar who will become a "moth". A chrysalis is a protective layer that appears when a caterpillar who will become a "butterfly" sheds its exoskeleton the final time.

Soon the little caterpillars find themselves wrapped snug and tight inside their warm bed called a chrysalis.

In the chrysalis stage, the caterpillar attaches itself to a branch for support and hangs upside-down. As it has done several times previously, one final time the exoskeleton layer of its skin splits open and is shed. This time, however, rather than a new layer of skin, the chrysalis appears with the caterpillar inside. The magical chrysalis transformation stage begins!

For ten long days the little caterpillars sleep peacefully in their chrysalis. As they sleep, butterflies from the nearby trees play without even noticing the sleeping little caterpillars.

The casual observer may think that because the chrysalis is motionless there is very little activity occurring during this "resting stage." However, within the chrysalis shell the caterpillar's entire structure is broken down and completely rearranged into the wings, body and legs of an adult butterfly. It is literally a total transformation!

One bright, sunny morning Bonnie calls out to her sleeping caterpillars, "It's time to wake up and get out of your chrysalis."

The chrysalis stage will last from 5 to 15 days, depending on the species. However, some species will last up to a year. During this phase the pupa does not feed, but instead gets its nutrients from the food it has eaten and stored during the caterpillar (larva) stage.

With sleepy eyes the little caterpillars squeeze out of their chrysalis'. Bonnie beams with pride as they try to open their eyes in the bright morning sunlight.

The **fourth and final stage** of the butterfly life cycle is the adult. Once the chrysalis casing splits, the butterfly emerges as an adult. Within days the new adult butterfly will mate and lay eggs to begin the cycle all over again. Sadly, most adult butterflies will live only a week or two.

The little caterpillars look at each other in disbelief at what they see! They are no longer little caterpillars; they have become beautiful butterflies, just as their mother had promised. They flutter their colorful wings as their bodies glisten in the sunshine. They are so happy!

A butterfly undergoes a miraculous process called "complete metamorphosis" during its life cycle. This means that the butterfly changes completely from its early larva stage as a caterpillar, until the final stage, when it becomes a beautiful and graceful adult butterfly.

While the butterflies from the other trees are out for their morning flight, the new little butterflies stretch their wings and take to the sky. As they flutter in the fresh morning air their mother calls out to them, "Be good, my beautiful *BUTTERFLIES!*"

They had become

JUST LIKE HER!

Butterfly Life Cycle

Stage One

Egg (Ovum)

Stage Two

Caterpillar (Larva)

Stage Three

Chrysalis (Pupa)

Stage Four

Butterfly (Adult)

Dedicated to Scarlett, my beautiful butterfly

The Lesson of the Butterfly

A man came upon a chrysalis in his garden. One day a small opening appeared in the chrysalis. He sat and watched for several hours as a butterfly struggled to force its body through the tiny opening in the chrysalis. The butterfly worked and struggled some more, and then it seemed to stop making progress. It appeared as if the butterfly had progressed as far as it could and it could go no further.

The man decided to help the little butterfly. He took a pair of scissors and cut a larger opening in the chrysalis. Then, with very little effort or struggle, the butterfly easily emerged. However, the man was saddened when he noticed that the butterfly had a large, swollen body and small, weak, shriveled wings.

The Lesson continued...

The man continued to watch the butterfly. He expected that, at any moment, the wings would enlarge and expand to be able to support the body, and the body would contract, allowing the butterfly to triumphantly take flight.

He waited and waited, but nothing happened. The butterfly spent the rest of its life crawling around on the ground with a swollen body and weak, shriveled wings. It was never able to fly. Never.

What the man did not understand was that the restricting chrysalis and the struggle required for the butterfly to get through the tiny opening would force fluid from the swollen body of the butterfly into its wings, causing the wings to expand so that it would be ready for flight once it achieved its freedom from the chrysalis.

Life is a butterfly.

www.ingramcontent.com/pod-product-compliance
Lightning Source LLC
Chambersburg PA
CBHW040453220526
45473CB00004B/1623